Emergency mobile irrigation system

DIY guide by Dino Rondic

Contents

Introduction 5

Principle of „EMIS" work 6

Parts.. 7

    1.   The Pickup truck 8

    2.   The water tank... 9

    3.   Electric water pump 10

    4.   Inflow and outflow hoses 11

    5.   Other parts 11

Assembly of EMIS 13

    a)   Pickup truck and the base of the water tank 13

    b)   Water Pump 16

    c)   Automatic water level controller..18

    d)   Hoses..18

Alternatives...20

Schematics...21

Check out my other titles 23

Introduction

What is *Emergency mobile irrigation system*?

„EMIS" (emergency mobile irrigation system) is a concept of small scale irrigation system used in emegencies as are unexpected droughts, lack of normal irrigation system due to the failures or unability of placing regular irrigation system on some terrain, and unexistance of regular irrigation system.

How „EMIS" works?

The principle of „EMIS" is very simple. A tank with water is mounted onto the truck capable of carying it. Tank is connected on one side with a water pump that is filling the tank with water from some source (river, lake...), and on the other side with an exit hose that can be connected with a system that delivers water to the plants (ex: drip irrigation), or it can be used for direct watering of the plants.

Since the truck is caryng the tank, you can drive to the nearest (or farest, it's you choice) water source, fill the tank and then return to the plants that are in need of water.

How much will it cost to build „EMIS"?

This system is generaly very cheap, depending of course on the price of the parts in the region you live in. Main parts are:

- Used pickup truck capable of caryng the tank
- Water tank
- Water Pump

I can not know, how much does it cost to purchase these parts where you live, but generaly these parts are easy to find and they are cheap.

Principle of „EMIS" work

After the „EMIS" is built, everything else is prety easy. Just by following these steps, you will be able to water your plants easily:

1. Drive the EMIS to the nearest, or cleanest water source
2. Turn the water pump on and let the pump fill the tank
3. After the tank if full, automatic water level controller will shut the pump off
4. Drive back to the plants and plug the exit hose into the water delivery system, or water the plant by you self

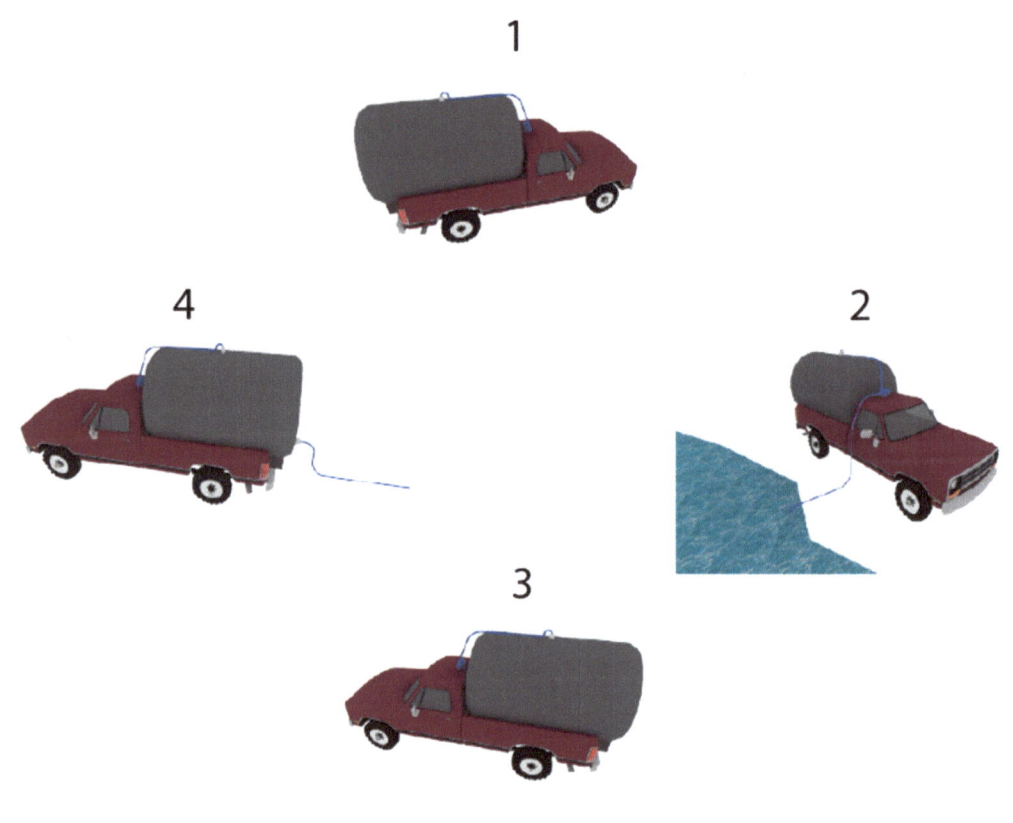

1 Illustration: Workflow

Parts

To build EMIS, you will need 5 main parts, pickup truck that is capable of caryng water tank, a water tank, electric powered 12v water pump and inflow and outfrow hoses. On the folowing illustration you can see these parts.

2 Illustration: Parts

1. A pickup truck capable of caryng water tank
2. Water tank
3. Electric powered water pump
4. Water inflow hose
5. Water outflow hose

Also some additional parts will be needed, as are automatic water level pump controller, base for the water tank, some valves and hose adapters, wires to connect the pump and the controller to the alternator and the battery, and a couple of switches.

Now, we can begin with detailed explanations of all of the parts. Lets start with the pickup truck.

1. The Pickup truck

Main characteristics of the truck that you need to follow when trying to find one, are that truck must be capable of caryng the water tank, so if you are planing to mount 5 ton water tank, be sure to ask the seller that truck is capable of caryng 5 tons of load. Next thing is that that truck needs to have place at the back for the water tank you are planing to mount. And finaly, besides working engine truck needs to have working alternator and the battery.

On the illustration below, you can see some pickup truck models that are suitable for the job.

3 Illustration: Some suitable pickup trucks

What type of pickup you will use (compact, mid-size, full size or hevy duty) , totaly depends on your needs, that is on the volume of the water tank you will mount.

2. The water tank

The water tank is mounted on the back of the pickup truck. The only thing you need to pay attention to is the volume of the tank (0.5, 1, 1.5 ... tons). Although on some previous illustrations tank is presented in the shape of the roller (drum) it will be easier and safer to mount rectangular shaped tank. Also, tank needs to have 1 entry and 1 exit point, for the water inflow and outflow hoses. On the outflow hose you will install a valve to controll the water outflow during watering and to stop ste water when you need it to be stoped. On the next illustration, you can see some types of water tanks you can mount onto the back of the pickup truck.

4 Illustration: Some suitable water tanks

3. Electric water pump

Electric water pump is used to fill up the water tank when you turn it on. The pump is directly connected to the truck's alternator, so when you are choosing the right pump for you, be sure to choose the pump with the puwer supply same as the output of the alternator (usually 12V DC). Since the alternator is giving DC voltage, buy the pump with a DC power supply. On the picture below, there are some suitable water pumps.

If your water tank is a bit larger, 12V pump will take some time filling it. If you are unsatisfied with the time your pump will take to fill the tank, consider buying 12V DC to 110/220 V AC power inverter, so you can install larger pump. You can connect inverter to the alternator and then the pump (with the 110 V AC or 220 V AC power supply) to the inverter. Remember that, after installing the pump, in order for the pump to work, car will have to be turned on so the alternator is running.

To calculate how much time will take the pump to fill the water tank, see how much that pump pumps water per minute (60l, 120l, 200l, 250l...), and divide the volume of the tank with that number. Ex: water pump pumps 60 liters per minute, tank has volume of 1000 liters, so 1000/60 = ~17 minutes).

12V DC water pumps

220V AC water pumps

4. Inflow and outflow hoses

Only thing here to vory about is that diameter of inflow hose is the same as the diameter of water pumps input, diameter of the hose that goes from the p0ump to the tank is the same as the diameter of water pump output, and the diameter of the outflow hose fittls to the valve on the exit point of the water tank. Regular garden hoses can be used as well as all other kinds.

Note that the hoses diameters can be increased or redused bu using hose adapters (reducers connectors).

5 Illustration: Some hose reducer connectors

5. Other parts

We will mention some other necessary parts here, but note that you might need more parts depending on specifications of your main parts and/or your needs.

a) Switches. You will need a couple of switches, to controll the work of the pump and the automatic water level pump controller. You can install those switches where ever you want. Some pumps ad vater level controllers have their own switches, if that is the case, you will not need additional switches.

b) Automatic water level controller. This device is controlling the pump's work. When you turn the pump on, you wont be able to see the level of the water in the tank, so this device will detect when the tank is full and it will turn the pump off when the water reaches top level. When you purchase this device, a sheme of how to connect it to the pump will come with it. Be aware that the voltage supply of the water level controller needs to be right so it would not burn. If the supply voltage of the battery is different from the voltage of the controller, ask for the advice from someone how to adjust the voltage of the battery to the voltage of the controller.

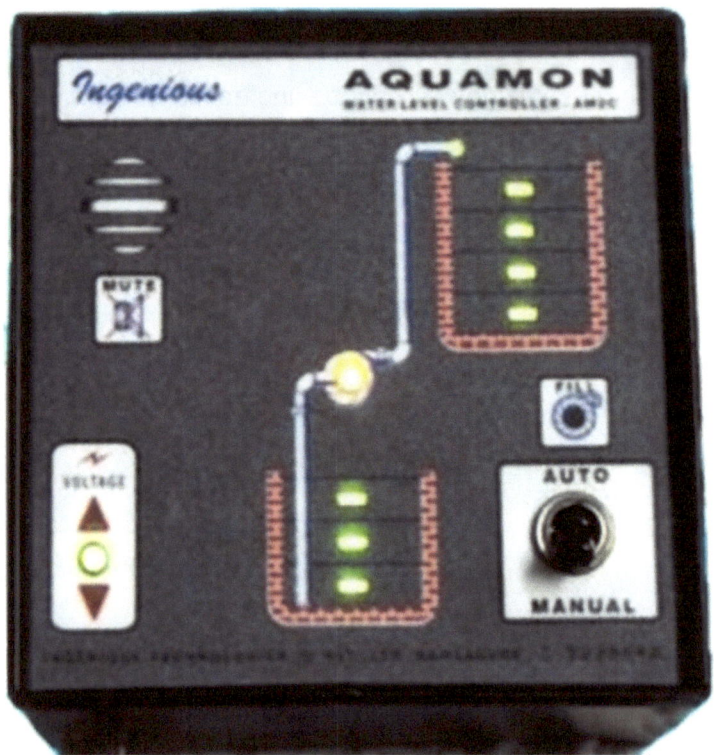

6 Illustration: an example water level controller

c) Wires. To connect the pump and the controller to the alternator and the battery, you will need some wire. You can use regular electric wires for this job.
d) The valve. At the exit point of the water tank, there needs to be installed the valve that will regulate the outflow of the water from the tank.

That is about it when it comes to the parts of the EMIS. Now we can continue to the assembly of the EMIS it self.

Assembly of EMIS

In this section we will put together all of the parts of the EMIS into one fully functional system.

a) Pickup truck and the base of the water tank

When you have found ideal pickup truck for the EMIS, all you have to do to prepare it for the water tank mounting is to make sure that the tank will fit on the back of the truck, and if the tank is not flat at the bottom to make the base for it so it would not move from side to side while the truck is moving.

Again, it is much easier to mount the water tank with flat base (ex: rectangular shaped water tank).

In the case where water tank is of rectangular shape, all you need to do is to place the tank on the back of the pickup and possibly use the bolts to secure the tank with the bed of the pickup if the tank has the extension for bolting to the base. You can also make the base for the rectangular shaped tanks, as shown on the illustration on the next page.

In the case where the thank is of the drum shape, you will need to make the base for the tank, and use bolts to secure that base to the pickup bed (that is if the tank has no abbility to stand still by it self.. The base will have to have the curved shape on the upper part in order for the tank to fit and it will have to have extension on the bottom in order for the base to be secured with the bolts to the pickup bed.

Note that before you place the tank, if it has any possibility to be secured to the bed with bolts, drill the holes into the bed so you can use bolts after you place the tank.

After you put the tank on its place, you will have to make sure it wont drop from the back of the pickup by placing the blocking metal bar and bolting it to the bed sides.

Step by step:
1. Make sure of what shape the water tank would be
2. Make sure if the tank has the abbility to be bolted to the bed
3. If the tank is drum shaped make the tank base as shown on illustration on the next page, and if the tank has flat bottom it is optional will you make the tank base or not.

4. Place the base (if needed) and secure it with bolts with the pickup bed.
5. Place the water tank after you have installed valve at the tank exit point.
6. Secure the tank with the metal bar.

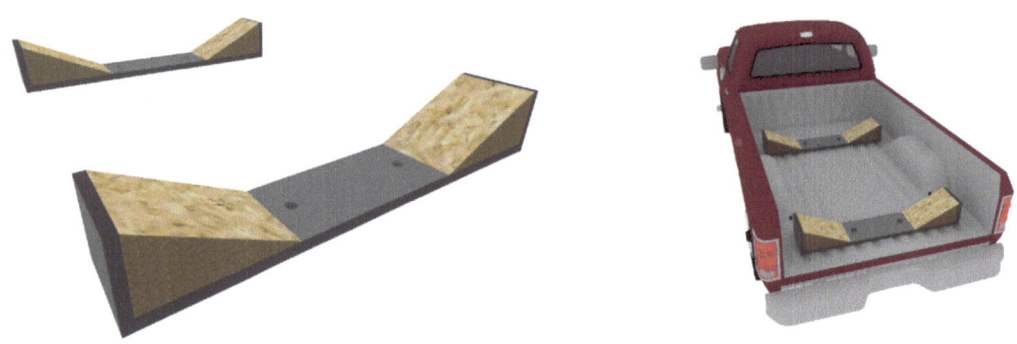

7 Illustration: Example tank base for the drum-shaped water tanks

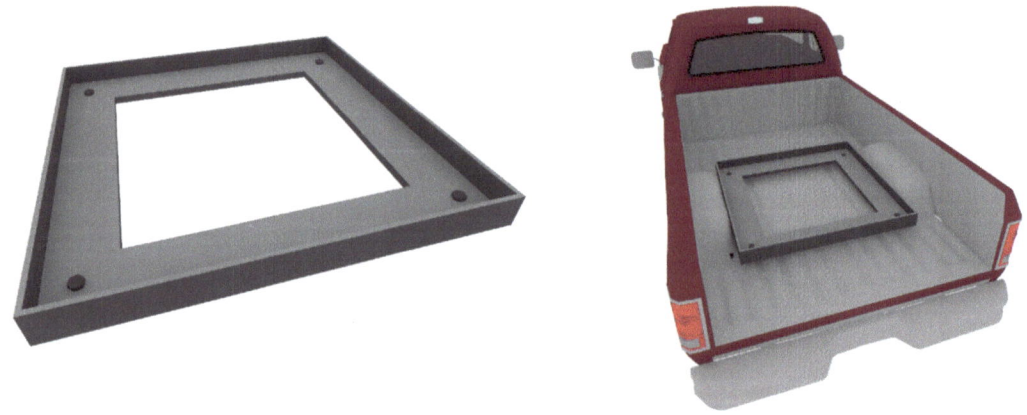

8 Illustration: Example tank base for the rectangular-shaped water tanks

Note that if the water tank has the extension at the bottom for bolts and screws you dont need the base, or if you secure the tank with the metal bar enough for the tank not to move you also dont need the base.

9 Illustration: Pickup truck with water tank mounted and secured

b) Water Pump

First of all, when it comes to the water pump type, you can use almost any type of pump you want, but electic powered pumps are best choice because you already have power source (alternator) and you can install automatic water level controller. If you can find 12V DC pump (or even 24V DC pump, depending on the output voltage of the alternator) it would definitely be the easiest choice concerning that you dont need to install additional devices as are inverter for the pumps powered by AC voltage or even DC voltage converters for the pumps that are powered by different voltage then the alternator output voltage.

However, in the case you want to install water pump that is powered by AC voltage, you will need to install power inverter. It is the device that converts 12 or 24 V DC into 110 or 220 V AC (depending on the input voltage value of the pump you have). You will connect the inverter directly to the output of the alternator, and then the pump to the inverter output. Choosing the AC over the DC powered pump will have the advantage of greater water flow per minute. But be aware that the pump's power must not be greater than the power of the alternator or else the pump will burn.

In the case where the pump is powered by DC voltage but the voltage is different than the output voltage of the alternator, you will need to use DC voltage converters to adjust the voltage of the pump with the voltage of the alternator.

The easiest way is to use 12 or 24 V DC water pump. That way you can connect the pump directly to the alternator, installing between the alternator and the pump, only the swich to controll the on/off function of the pump.

Location of the pump is not very important. Easiest way is to place and secure the pump right next to the water tank on the pickup bed.

On the next page you can see shematics of the pump connected to the power source. If you cannot find the inverter or the DC voltage converter, google some DIY circuits, and you will find plenty of solutions to make these devices by you self.

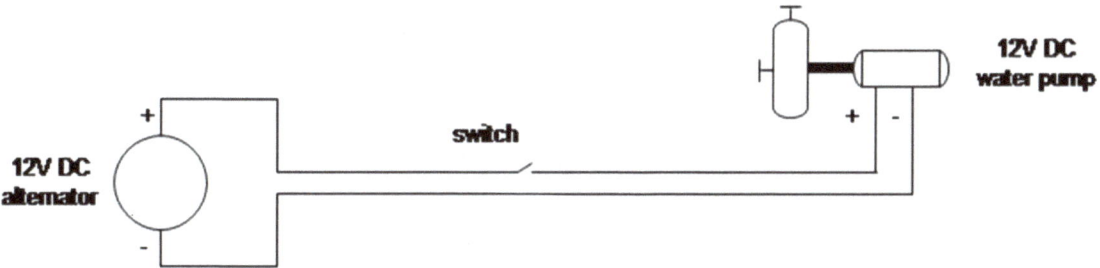

10 Scheme: 12V DC water pump

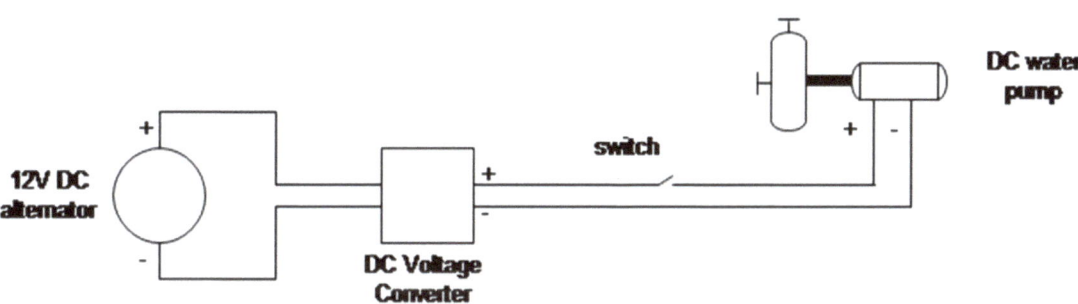

11 Scheme: DC water pump of different voltage than alternator

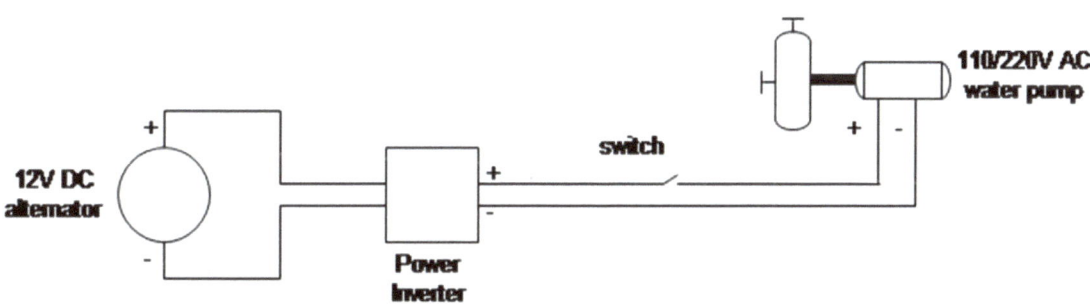

12 Scheme: AC water pump to DC power source + power inverter

c) Automatic water level controller

This device detects the water level in the water tank and turns on/off the pump when the water level is min/max. It is usualy powered by 12V DC or 24V DC, so you can connect it to the car battery directly since it does not require much energy to run. When connecting it, dont forget to install the switch if the controller does not have one built in.

It would be irresponcible from me to provide you with the connection diagram, since there exists many types of controllers, with variety of connecting possibilities. When you purchase the controller, the diagram will come with it, or if you are making your own DIY controller circuit, everything is explained well on the circuit schematics.

However, if you cannot find the controller to purchase or you simply want to build one by you self, again, google „watter level controller diy circuit diagram", and there will be a lot of suitable results. I will provide some circuit diagram, but there is so much different designs that requre different parts that are not widely available for purchase but can be ascuired from old circuits and devices, so it is up to you to choose the circuit design for which you can find parts easily, just google „diy water level controller circuit".

Location of the controller is up to you, but it will be safest to install it inside of the car cabin where is safe from water and dust.

d) Hoses

This is prety simple task to install these hoses. First hose connects the water source with the pump, second one leads from the pump into the water tank, and the third one leads from water tank to the where ever you need it to go. Only problem could be attaching the hoces to the pump and the water tank since there can be some differencies of the diameters, but that can be simply solved by using the reducer connectors.

Since we have explained all of the parts and their functions, now simply assemble the EMIS by following the steps form **a** to **d**. To go through these steps in short:

1. Prepare the pickup truck for water tank maounting (make the base if needed).
2. Mount the water tank onto the bed of the pickup truck and install the valve at the exit point of the tank.

3. Install the water pump also on the pickup bed and connect it to the alternator.
4. Install the water level controller and connect it to the battery, tank and the pump.
5. Connect the pump, and the tank with the hoses, and leave one hose for the water source and one for the water outflow.

That is about it. See the illustration and schematics below of the finished EMIS.

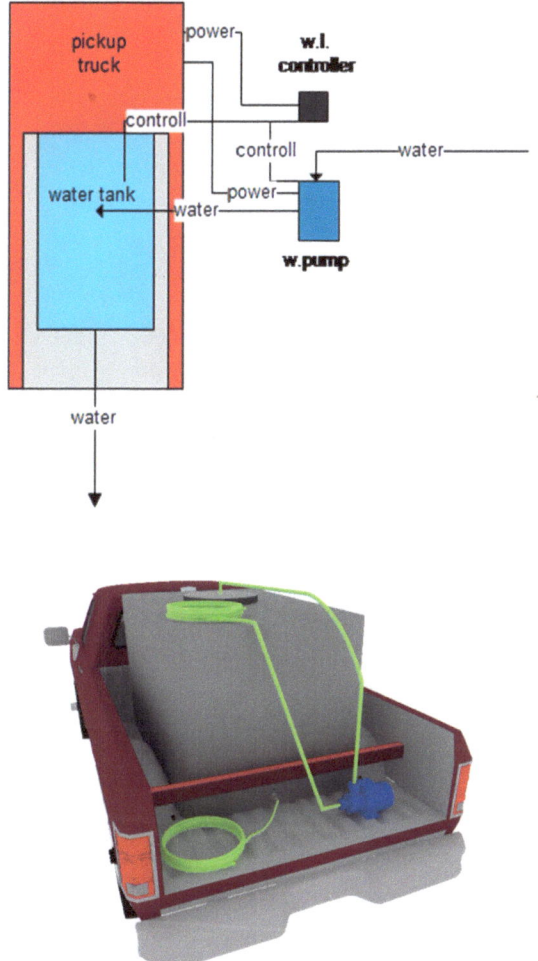

13 Illustration: Finished EMIS

Alternatives

In the case you need bigger water tank, the same principle of assembling EMIS can be applied to larger truck capable of carryng your water tank.

Also, this system can be built using a regular car and a trailer big enough to fit tra water tank on it.

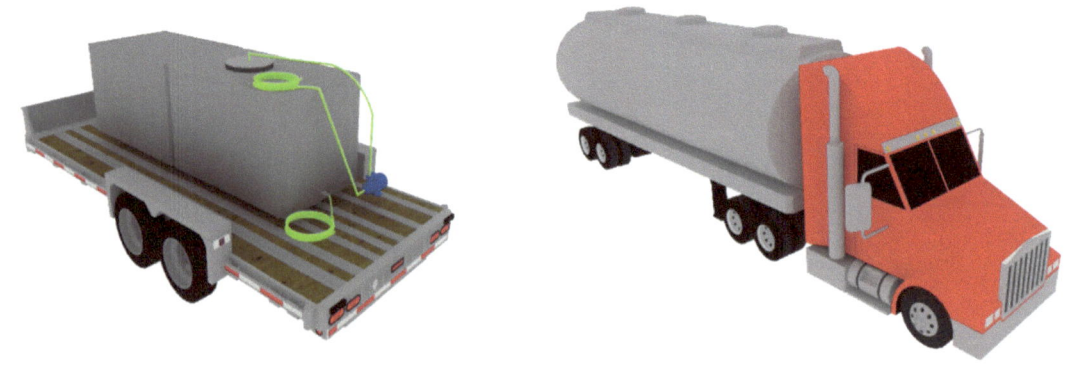

14 Illustration: Pickup alternatives

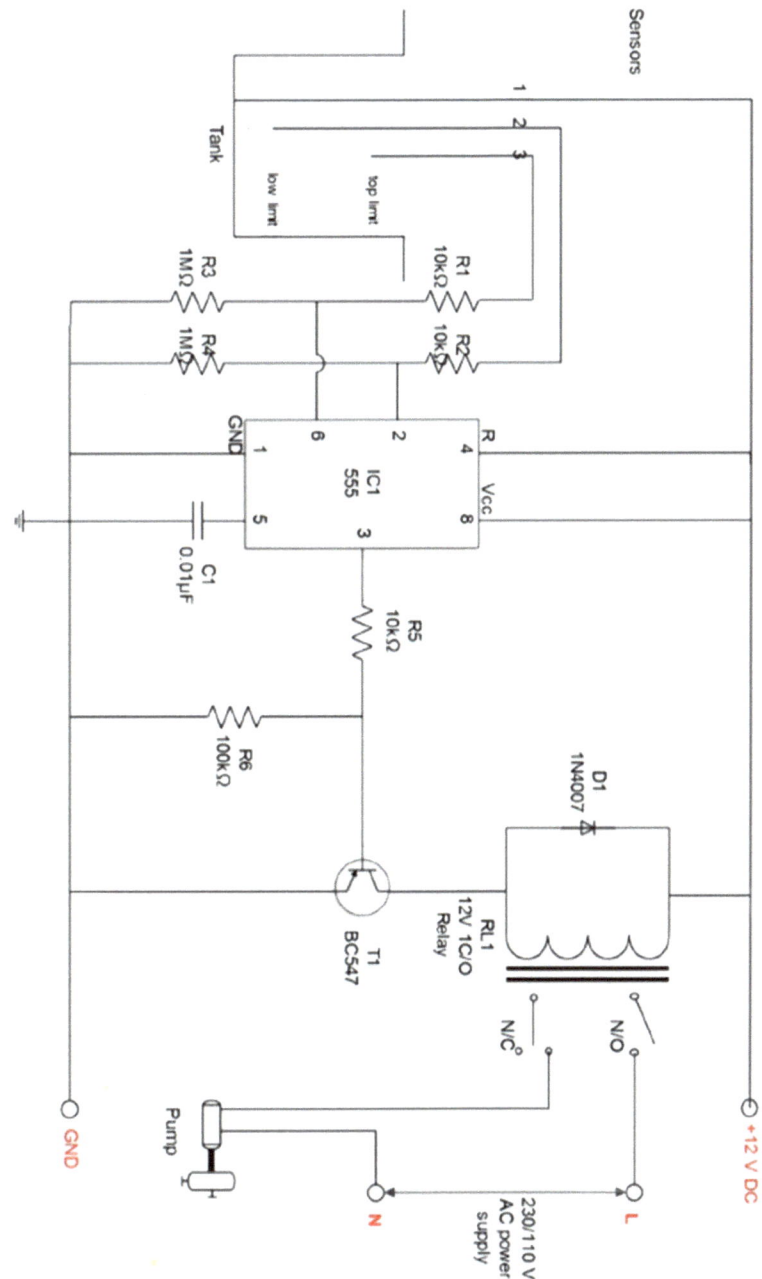

Diagram: Water level Controller

*Sensors are regular copper wires

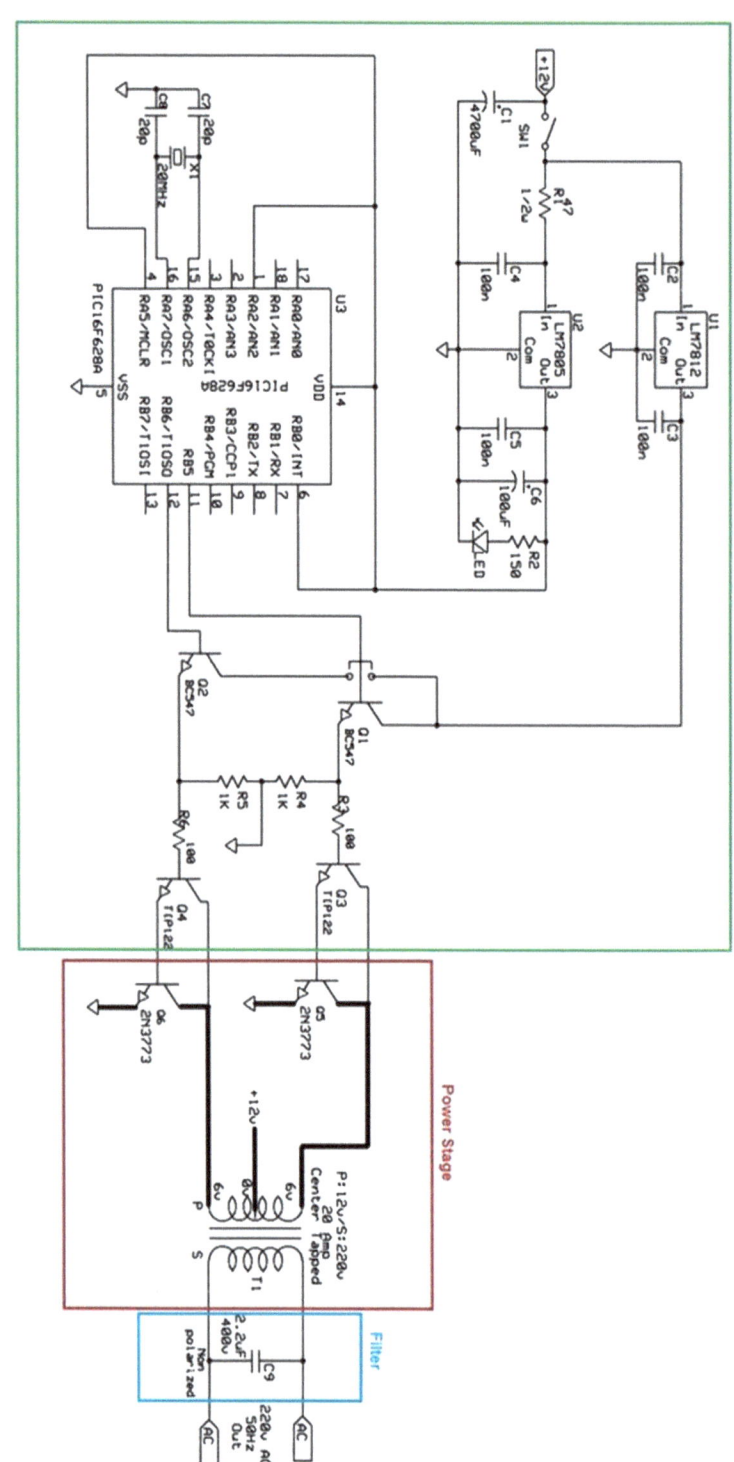

Scheme: 100 to 5000w - 12vDC to 220vAC Power inverter

Check out my other titles

Comprehensive guide to building Greenhouses

Greenhouses are sources of continuous throughout-a-year food production. They provide ideal conditions for growing plants for your own or commercial use.

This book is a guide that shows how to plan, organize resources and build your own greenhouse. It contains a step by step illustrated guide for building small size cold frames, small to large scale PVC greenhouses and glasshouses.

Detailed guide on making solar water heaters

Detailed guide on making cheap yet quality PVC solar water heaters for the purpose of harvesting sun power for free summer-time hot water in use your home or cabin. This book is a guide to making one and the best type of homemade solar water heater.

It is cheap, quality and will reduce your electricity or gas bill significantly.

Homemade Accumulating irrigation system for medium scale gardens, growing fields and small to large scale greenhouses

This is complete illustrated guide showing you how to make Accumulating Irrigation system on your own from scratch.

This system is simple, reliable, and above all, much less expensive than purchasing commercial system of a same purpose. This irrigation system, after built, can be completely controlled with one electrical switch from your house.

The guide contains: making water pump with water level switch controller, water tanks and delivery systems. "AIS" is suitable for growing fields and for greenhouses.